HEART OF WINTER

POEMS & TRANSLATIONS

BASED UPON THE WRITINGS OF

KNUD RASMUSSEN & THORILD WULFF

AUTUMN RICHARDSON

XYLEM BOOKS 2018

xy01 The Look Away (2018)
xy02 Field Notes, Volume One (2018)
xy03 Heart of Winter (2018)

Heart of Winter

Copyright © Autumn Richardson 2013–2016
The author's moral rights have been asserted

First published in 2016 by Corbel Stone Press
This edition copyright © Xylem Books 2018

ISBN: 978-1-9999718-2-3

Cover image:
Knud Rasmussen (1879–1933)
Courtesy of the Library of Congress,
Prints & Photographs Division, [LC-B2-556-2]

Leading image:
Flora Danica, Fasicle 41
G.C. Oeder (1761–1883)

Xylem Books is an imprint of Corbel Stone Press

Preface	xi
Across the Inland-Ice	1
Four Inuit Songs	31
The Last Diary Entries of Dr. Thorild Wulff	45
Epilogue	67
Botanical Index	75
Acknowledgements	83

PREVIOUS PAGE

Luzula nivalis

Flora Danica, Fasicle 50

G.C. Oeder (1761–1883)

PREFACE

In 1916–17 Greenlandic-Danish ethnographer and polar explorer Knud Rasmussen (1879–1933), along with six others, formed the *Second Thule Expedition*, their purpose to chart a little-known area of the far north-western coast of Greenland – 'the last unknown reach of Greenland's north coast on the stretch between St. George Fjord and de Long Fjord.'

The full party comprised:

Dr. Thorild Wulff	*botanist and biologist*
Lauge Koch	*geologist and cartographer*
Hendrik Olsen	
Ajako	
Nasaitsordluarsuk	*called Bosun*
Inukitsoq	*called Harrigan*
Knud Rasmussen	*chief and ethnographer to the expedition*

Sadly, both Olsen and Wulff did not return from the expedition; Olsen disappeared and Wulff succumbed to malnutrition and exhaustion, dying only days before rescuers arrived.

~

Heart of Winter is a collection of found poems assembled from the expedition journals of Rasmussen and the last diary entries of Wulff, as published in *Greenland by the Polar Sea* (Heinemann, London, 1921). These poems are accompanied by a series of new English translations of Inuit songs from Greenland and North America, first transcribed into Danish by Rasmussen and published as *Snyhytten Sange* (*Snow-hut*

Songs, 1930). Wherever possible I have retained the spellings, punctuation and botanical abbreviations from the original texts.

~

During the process of writing *Heart of Winter*, I came to realise that I was, in part, answering a desire that has roots in my childhood; a longing to explore and understand landscapes of the north. I grew up in a place, which, to my child's mind, was both measureless and deeply mysterious: the vast expanse of boreal forest, punctuated with countless rivers and lakes, which stretches across the Canadian Shield in northern Ontario, Canada. My interest in indigenous cultures also began there, through my early readings of, and listenings to, Ojibwe myths – stories of *manitous* and shapeshifters, of sentient animals, of visions granted to those who sought the wildest, bleakest places. Soon afterwards I discovered the adventurous writings of environmentalist Farley Mowat. I read his first book *People of the Deer*, documenting the lives of the *Inhalmuit*, the Inuit peoples inhabiting the barren lands west of Hudson Bay, at the age of nine or ten, and from that moment became obsessed with going further north – of one day witnessing the great caribou migrations, of travelling through the vast tundra, and onwards to the furthest ice-covered shores of the most northerly outposts, and of meeting the peoples, the animals and the plants, that made their homes there.

~

It was natural that my passion for northern cultures would eventually lead me to the life and work of Knud Rasmussen.

Himself partially Inuit, he was born in what is now Ilulissat, Greenland, to an Inuit-Danish mother and Danish father. He was a speaker of Kalaallisut – West Greenlandic, the standard dialect of the Greenlandic Inuit language – and from his earliest childhood he learned to hunt traditionally, to drive sled dogs and to journey and survive in Arctic conditions. Perhaps it is because of this upbringing that his style of exploration was so sensitive to both the landscapes he traversed and to the peoples that he met along the way.

I have read many accounts of exploration, but what shines uniquely through Rasmussen's writing is his genuine connection to the environments he is passing through; a detailed, warm engagement with place. He was an ethnologist, an explorer and a writer, but it became abundantly clear to me as I kept reading his words, in various translations, that he was also a poet. As I researched further and found a copy *Snehyttens Sange*, I began to understand even more deeply where the beauty of his language may have originated.

~

As I had no understanding of Danish at the beginning of this endeavour, it was a painstaking labour of love to translate the poems from *Snehyttens Sange* – but as the words revealed themselves I was taken aback by the vividness, the clarity and the vitality that chimed through them. A startling immediacy leapt out of these words, now nearly a century old – and in some instances, possibly much older.

It was undeniable that the people who composed these words were at one with the place in which they lived. Their landscape was not only a food source, nor just a tool; they were not engaging with it in a purely utilitarian way. Within these poems of everyday life there is *joy*; joy in a change in the weather, joy in receiving food, joy through gathering with others in song. I was so moved by this – this clear, focused gratitude that was carried through the words, and by the spontaneity and purity that radiated from them. And yet I also felt bereft, realising that this joy, this gratitude, is now rarely a part of modern life. Most of us in the west don't live on the knife-edge of survival. As members of industrialised societies the majority of us have, perhaps without realising it, traded vitality for security; once finely honed senses for material ease. We have opted for human company rather than communion with the myriad other lives surrounding us.

~

The writings of Rasmussen, the Inuit songs, and the final notes of Dr. Wulff, share moments of life at its slimmest, its most immediate. Each recount life-threatening and life-sustaining experiences, and the endurance of incredible hardship, yet in their words there is also at times a euphoria; an awareness of the immense beauty surrounding them, and a recognition of the preciousness of each small life that they encounter.

This immediacy, this clear chiming presence within the experiences being shared, birthed a sense of longing in me, a small grief – for it is this presence, this vividness, that I feel is missing from much contemporary experience. For such vitality as this – a sharp-sensed, antennae-fine awareness – can only emerge from a genuine, deeply immersive relationship with life – all of life – and with death.

And so I embarked on this work, sinking into the words, and the worlds, of Rasmussen and Wulff, of Inuit women and men, sharing how they felt and saw and sang their landscapes. It has been a remarkable journey. And a privilege.

Heart of Winter is dedicated to the memory of Dr. Thorild Wulff and Hendrik Olsen. It is also dedicated to those intrepid sled dogs who transported both men and equipment, and to those who were hunted – musk-ox, seal, ptarmigan, reindeer and hare, amongst others – without whom no man would have survived. And finally, it is dedicated to those indiginous inhabitants, the Inuit, who shared their wisdom, their skills and their songs so generously.

Autumn Richardson
Fife, Scotland
January, 2016

HEART OF WINTER

PREVIOUS PAGE

Draba nivalis

Flora Danica, Fasicle 41

G.C. Oeder (1761–1883)

ACROSS THE INLAND-ICE

No sign of life
not a bird, not a plant.
Only lichens clothing
the sharp stones with grey.

A patch as big as a penny
can be more than a century old
here in this place
where vegetation is at rest
350 days of the year.

Deriving nourishment
chiefly from stone

the lichen is thus a plant
which in all its poverty
has eternity before it.

Dead calm, a clear sky,
temperature of 1° c.

Seven hundred metres above sea-level.

Here before us lies a depressing
and barren land
where the deep silence
is unbroken even by a bird
or the low murmur of water.

A hill sloping towards the sun.

Temperature of the air in the shade:
minus 11'8° c.

The thermometer with its ball of mercury:

1. On a light-brown, sunny cliff
of sandstone: minus 1° c.

2. On a sunny clump of saxifrage:
plus 2'8° c.

3. In cespitous moss: plus 9'2° c.

During the long march yesterday
we were quite suddenly
surprised by the visit of a young gull:

a storm blew it in here
and it was unable to find its way
back to the sea.

For a long distance
it fluttered feebly—
until the wind seized it—

carrying it further in
towards the wastes
and death.

The land is dry and even to walk on
but barren and naked as a desert;
a karst landscape which sacrifices
all water to its depths.

We've found the jaw-bone of a musk-ox,
and near to this, a fossilized piece
of an octopus from the Silurian period.

The vegetation is accordingly;
a few poppies, small stunted grasses,
mosses and lichens—but no animal life.

Only a lone wolf had long ago
left his imprint in the clay.

The first dog fell to-day.

We had no choice
but to divide it up as food
for the other dogs.

We do not attempt to hide the fact
that the very difficult conditions
have weakened us; our faces show
we have become very thin.

Of real provisions
we possess merely enough
for six days.

The transport was difficult
and of long duration,
partly across firns, partly across
snow-drifts, but today
we got out of the cauldron.

On both sides
high mountains fall
steeply
downwards
into a valley of snow.

Yet we've found great,
beautiful branches of coral
etched into stone—
signifying that even here,
within this heart of winter,
waves of a living ocean once lapped.

And thus, whilst century follows
century, there is only change.

A slope to the west. Calm, clear sunshine.

Temperature of the air in the shade:
plus 5° c.

The thermometer with its ball of mercury:

1. On a sunny, flowering tuft of saxifrage:
    plus 21'1° c.

2. 2 cms. down in dry, sandy soil:
    plus 14'2° c.

3. 6 cms. down in moist, sandy soil:
    plus 12'5° c.

4. 12 cms. down in moist, sandy soil:
    plus 7'3° c.

On this stretch of coast
Wulff has found a living saxifrage
with fully developed flowers
on stems an inch high.

In full bloom
its tissues are full of life—
although the temperature
of the air is minus 11° c
and there has of yet been
no thaw.

It has hibernated in perfect
stillness throughout the winter,
and now calmly continues its life,
as spring and sunshine dissolve
the ice which once surrounded it.

Thick whiteness, punishing winds.

We are forced to hibernate like bears
while the storm lashes around us.
An exhausted dog is lying out
in the drifts, whimpering pitifully.

Our last cup of coffee
and one portion but one left
of pemmican gruel.

The entire journey has been hampered
by crevasses which are difficult
to notice in the hazy atmosphere.

The weather is continually disturbed.
We've had to kill another dog.

PREVIOUS PAGE

Salix arctica

Icones Plantarum Novarum, Vol. 5

C.F. von Ledebour (1834)

In the midst of fog and hopelessness
we see a sign of life from land—
a small fly buzzes past us.

The glacier has been very porous,
with large, pointed ice crystals
and deep, round Cryokonite holes.

Through a dark bank of fog we reach
a large, very beautiful glacier lake.

We are now some 30 kilometres away
from 'the great land without mountains'
and we hope here our difficult experience
will have an end.

Thinking of tallow and reindeer meat.

Wulff's increasing exhaustion
is a source of anxiety to us;
the expression in his eyes becomes
weaker and weaker.

If only the incessant fog would diminish.

We have reached land—
we have returned to food and to life!

We have escaped from the terrible
embrace of the inland-ice.

Dr. Wulff has declared he cannot
continue, he must rest.
Koch is of the same opinion.

It is agreed that Ajako and I must
go to Etah for relief—Harrigan
and Bosun will remain in order
to hunt for Wulff and Koch,
who no longer have the strength
to pursue game.

Ajako and I set out on our walk.
We bring merely the strictly
necessary things—our kamiks,
my diaries, and nothing else.

We part from our comrades
in the best of spirits after a feast
of newly shot hares.

Dr. Wulff has made for himself
a comfortable little sleeping-place
on a moss-clad shelf.
Smiling, he waves good-bye.

May Ajako and I have
the strength we need to find people
and the speedy relief
that is needed for our comrades.

Kamik – a soft boot, traditionally made of sealskin or reindeer skin, worn by Arctic aboriginal peoples.

A laborious climb
up the mountain-sides
to a stony and cleft place
intersected by a number
of great rivers.

Small crevasses and more frozen
rivers. Some large lakes.

Below us stretch the skeletons
of two musk-oxen.

The fertile banks alongside the lakes
must surely house the prey we seek—

in vain we examine with the field-glasses
all cloughs, river-beds, and valleys—

but not a living form do we discover.

We spread our sleeping-skins across
oblong hollows filled with *Cassiope*.

The uncanny rush of rivers is heard
all around us. Weariness overcomes
my watchfulness. I must sleep.

The wind appears to be the only
guest in these harsh tracts
where even snow cannot rest.

But wherever small, clough-like
hollows give shelter from the snow

or where a river winds its way forth
from one of the countless lakes,

here grasses and willows flourish—
and upon these hare and musk-ox
survive

and upon them, the polar wolves
and the small fox, and the ermine
in its winter coat—

and so through each, life is brought
to this seemingly desolate place.

Summer has passed its climax.

Coolness has announced itself
before the cold, as have colours before
the snow—a last blaze-up before
the sleep of winter.

On thick, soft moss we walk along
the small mountain-rimmed lakes,
which wink at us like black, deep eyes.

Plains of grass, little lakes,
fresh tracks of reindeer.

The weather is raw and foggy
and I am faint with hunger.

We are faced with a river,
foaming huge and white between
enormous stones.

Without hesitation Ajako and I
seize each other's hands,
and thus propping one another up,
we go out into the water.

We come out on a plain stretching
widely and openly ahead,
with little rivers and vigorous
grass-meadows shining sun-gilt
against dark crimson stone-heaps.

We've caught three young hares—
food for the stomach, marrow
for the bones.

Once more a huge fire
flares up in the gloaming;
we will make blood soup—
it will give us warmth for the night.

We have travelled upwards
of 100 kilometres
since leaving our comrades.

It is one o'clock
when we lie down
to rest after fifteen hours'
walk without stop.

Every time I sleep I dream
about my home.

They saw us through the windows
and poured out—men, women
and children, like lava from a volcanic
eruption, overwhelming us
with loud shouts of welcome.

Hearty words and laughter resonated
in our ears, questions rained over us,
and it was as if big waves beat together
above our heads and swallowed us.

We have stepped across the threshold
from death to life, from the great silent
wastes into joy.

Towards the west
the living sea
unclosed by the dead quiet
of the polar-ice;
the smell of salt water
and pungent seaweed.

The wind calms
with the sinking of the sun.
Dusk throws its sharp shadows
across the earth
whilst the ocean gleams silver
between ice-mountains
and drifting floes.

In these fjord waters scents
of walrus, narwhal, and seals—
all those who will now return
us to life.

Beautiful ocean—I recognize you—
I am home.

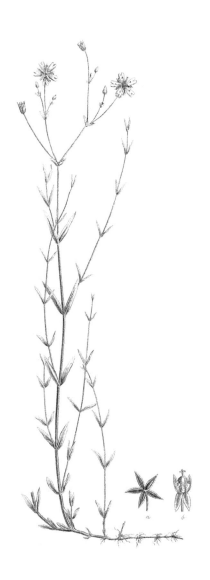

PREVIOUS PAGE

Stellaria longipes

Icones Plantarum Novarum, Vol. 5

C.F. von Ledebour (1834)

FOUR INUIT SONGS

These Inuit songs were originally collected and translated into Danish by Knud Rasmussen, and published in *Snyhytten Sange*, 1930.

BEVÆGELSE

Det store hav
bringer mig i bevægelse!
Det store hav
sætter mig i drift!
Det bevæger mig
som alger på stene
i rindende elvvand.

Himlens hvælv
bringer mig i bevægelse!
Det vældige vejr
blæser igennem mit sind.
Det river mig med,
så jeg skælver af glæde.

Uvavnuk. Kvinde,
Iglulik-Folket, Iglulik.

MOVEMENT

The great sea moves me

the vast waters
have sent me adrift

they move me so I tremble
like a weed in the currents.

The vast sky opens me

its immensity
blows through me
like a wind.

It tears through me
so I tremble with joy.

Uvavnuk. Woman,
Iglulik People, Iglulik.

EN LILLE SANG

Jeg synger en lille sang,
en andens slidte, lille sang,
men jeg synger den som min egen,
min egen kære, lille sang.
Og således leger jeg
med denne slidte, lille sang,
idet jeg fornyer den.

Søndre Upernivik,
Vest-Grønland.

A LITTLE SONG

I sing a little song,
someone else's worn,
little song,
but I sing it as my own;
my own dear little song.

And so I play
this worn out,
little song
and I renew it.

Southern Upernivik,
West Greenland.

SULT

Der var angst over mig—
I mit lille hus
holdt jeg ikke ud at være.

Sulten og forkommen
vakled jeg derfor ind over land,
uafbrudt snublende forover.

Ved 'den lille Moskusokse-sø'
holdt ørrederne mig for nar.
Jeg fik intet bid.

Videre sled jeg mig så
til 'det unge menneskes bredning'—
her havde jeg før fanget laks.

Jeg ville så gerne se
svømmende ren eller fisk i en sø.
Den fryd var mit eneste ønske.

Min tanke endte i sletingenting.
Den var som en line,
der helt løber ud.

Skulle jeg mon aldrig
få fast grund at stå på?
Trylleord mumled jeg hele vejen.

Kingmerut. Mand, Ellis River,
Queen Maud's Sea.

HUNGER

There was anxiety over me
in my little house;
I was not able to maintain myself.

Hungry and exhausted
I wavered over the land, stumbling,
searching for sustenance.

By 'the little musk-ox lake'
the trout evaded me. I got no bites.

I tired myself travelling
to 'the young man's river'
where once I had caught salmon.

Seeing their shapes in the clear
water below was my only wish.

My wishes came to nothing—
they were like a line I pulled up empty.

Will I ever find what I need?
Magic words I murmur, murmur.

Kingmerut. Man. Ellis River,
Queen Maud's Sea.

SANG TIL FORÅRET

Jeg var ude i kajak
og søgte land,

her kom jeg til en snedrive,
der var begyndt at smelte.

Så vidste jeg, at det var forår,
og at vi havde overlevet vinteren.

Og jeg blev angst for, at mine øjne
skulle være for svage,
alt for svage
til at se alt det dejlige.

Angmagssalik,
Øst-Grønland.

SONG OF SPRING

I was out in a kayak
seeking land.

I came to a snowdrift
which had begun to melt—

and so I knew that it was spring
and we had survived the winter.

And I had feared my eyes
would be too weak,

too weak
to see all of the beauty.

Angmagssalik,
East Greenland.

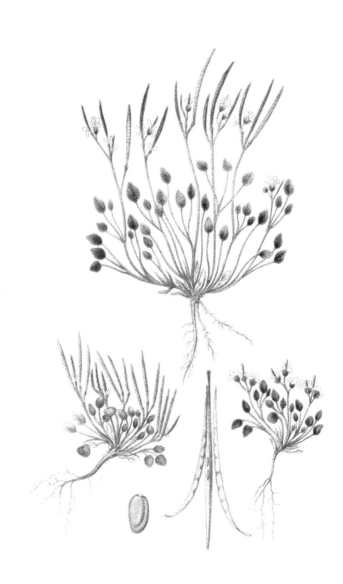

PREVIOUS PAGE

Cardamine bellidifolia

Bilder ur Nordens Flora, Vol. 1

C.A.M. Lindman (1922–1926)

THE LAST DIARY ENTRIES

OF DR. THORILD WULFF

I.

August 24th.

We start from camp at 9:15 A. M.
Descent rather steep.
Several glacier torrents are crossed.

Dead tired, half unconscious,
we reach the gneiss cliffs at 7:30 P. M.
after exactly three weeks' march,
four hundred kilometres
across the inland-ice.

The last dog is killed.
Tracks of reindeer and hare.

II.

The edge of the inland-ice, 8 P. M.
Calm. Fog. Drizzle.

We lie down to sleep
on mountain shelves.
Cold, tent cannot be pitched.

L. leucopterus. Vegetation autumnal.
5° during the night, hoar-frost.

Salix arctica quite light yellow.
Luz. confusa., *Sax. oppositifol.*,
Cernua nivalis, *tricuspidata*, the latter
vigorous, still in bloom, blood-red.

Papaver, *Draba*.

L. leucopterus – likely *Larus leucopterus* (now *Larus glaucoides*); Iceland gull.

III.

August 25th.

Boiled hare. Delicious liver
and heart. Strong soup, but I am ill
from the meat diet.
Thinking merely of peas, jam, bread,
fruit, brandy, coffee, chocolate.

Nevertheless I eat as much as I can
to regain the wish to live,
and to conquer my weakness.

New ice formed last night on the lake.
Feel continually reduced in strength.

Cassiope, Stell. longipes, Aspid. fragrans.

IV.

Three more hares caught, all young
with grey heads. One eaten raw, two boiled.

Potentilla nivea, rubricaulis, emarginata,
Dryas, broad-leaved, smooth, octopetala-like.
Pedicularis hirsuta.

Very commonly *Myrtillus uliginosa,*
scattered extensive mats, and *Salix arctica,*
with broadly oval and narrow
lance-shaped leaves.

V.

Knud and Ajako started out on foot
for Etah, approx. 200 kilometres,
the straight road across land to send us
relief sledges and provisions.

Myrtillus uliginosa, Pyrola uniflora,
Wahlbergella (large, not triflor.).

Drink warm water for supper.

See page 16. After some rest, Wulff, Koch, Bosun and Harrigan continued slowly
westwards, with the hope of meeting the relief party on its way back to rescue them.

VI.

August 26th.

I am sleepless. Clear cold night.
Eat in the morning the last remnants
of the last dog.

VII.

We continue westward.
Thrown away theolodite, two cameras,
bandages, clothes; everything which
we can yet do without.

To think of collecting plants is impossible.

We do not even carry tent or Primus,
merely guns, as we are dead-tired.

Remains now the most serious flight for life.
I am only a skeleton and shiver with cold.

This will be a march towards death
if a miracle does not happen.

Plants and films and notebooks remain
by the edge of the inland-ice under a stone
where we slept the last two days.

VIII.

Lesquerella, Hesperis, Cerast. alp.,
Kobresia, C. nard., Erioph. polyst.

Poa cenisia, Trisetum, Heirochloa,
Luzula nivalis, Sax. opposit. Flower.
Alsine verna, Silene acaulis.

PREVIOUS PAGE

Poa cenisia

Flora Danica, Fasicle 43

G.C. Oeder (1761–1883)

IX.

Surely the relief sledges will reach us
soon and save us from this struggle
with death which has lasted
since the middle of May.

When deadly indifference to life appears
and weakness gets the upper hand,
even food phantasies disappear
and the thoughts occupy themselves
with those at home, and with the strange
sum total of life.

X.

Heavy, stony, cliff-terrain.
In the evening a young hare.
Minus 1'4°.

Camp for the night upon moss
near a small border-lake.

Day's march approx. 6 kilometres.

XI.

August 27th.

To sleep 11 P. M. on the mossy slope.
Fog, minus 0'5°, a little snow.

4 P. M. The hunters return—glorious—four
hares in one day for four men—this means
life for us. Entrails eaten raw. Blood goes
into the soup.

My strength, which was almost exhausted,
returns. These last four days I have
been nearer to death than life.

Fire of *Cassiope* and old dry branches
of *Salix arctica*, finger-thick.

Vegetation finished for the year, everything
yellow and brown, ready for winter's rest.

Fruit of *Cassiope, Sax. opposit., tricusp.,
Dryas, Potentilla, Drabae, Wahlberg. affinis*
and *triflorum*.

XII.

A loon, geese, terns, buntings
in flocks.

Midnight gloom, gneiss knolls.

XIII.

August 28th.

Cold. Fog. Falling Snow.
Diarrhoea. Misery.
Start 1 P. M. through thick snow.

Cystopteris (com.), Lycopod. Selago.

Rich plankton in several little lakes.
Red-polls in flocks, terns, falcons,
plenty of animal life.

Sax. cernua, foot high with top leaf.
Myrtillus ulig., blood-red, very common,
though always without fruit.

J. biglumis, Epilob. latifol.,
Oxyria, Draba nivalis, hirta,
Cardam. Bellidifol.

Bosun catches a young hare.
Driving snow, fog.
Shared the entrails at once,
ate them raw, warmth in the body.

Strenuous march until 12:30 A. M.
without finding more game.

XIV.

August 29th.

I am half-dead, but found *Woodsia ilvensis*.
Lay down at 7 P. M. for I will not hamper
the movements of my comrades upon
which hangs their salvation.

PREVIOUS PAGE

Cardamine bellidifolia

Icones Plantarum Novarum, Vol. 3

C.F. von Ledebour (1831)

EPILOGUE

September 10th.—Wulff is dead. This evening the relief sledges returned with Koch, Harrigan and Bosun.

It was ordained, then, that after all he should not have the strength to continue, but must give up just as he had reached land and was not far from men. This last death takes me absolutely by surprise. I know that he was exhausted, but so were we all; that death was approaching when Ajako and I departed I did not suspect.

What a tragic death, just as he had toiled through all dangers and seemed safe at last. I cannot understand it—I cannot understand it!

Yet it is true; the man with whom for a long time I have shared good and evil I shall see no more. Like his sledge comrade Hendrik, he has entered the great peace.

Knud Rasmussen's journal, 1917

PREVIOUS PAGE

Alsine verna

Florae Austriaceae, Vol. 5

N.J. von Jacquin (1778)

BOTANICAL INDEX

Alsine verna : now *Minuartia verna*; Spring sandwort. The Alsine genus is now obsolete and has been replaced with Minuartia – plants commonly known as sandworts or stitchworts.

Aspidium fragrans : current name *Dryopteris fragrans*; Fragrant woodfern.

Cardamine bellidifolia : Alpine bitter-cress.

Carex nardina : Spike sedge.

Cassiope : *Cassiope tetragona*; Arctic bell-heather, Arctic cassiope.

Cerastium alpinum : Alpine mouse-ear, Alpine chickweed.

Cernua nivalis, tricuspidata : an anomaly in Wulff's notes. In botanical nomenclature the first part of the binomial references genus, the second part the species. 'Cernua' is used adjectively as the second part of a plant binomial to mean 'nodding' or 'drooping'. It is possible the first part of the Latin name was simply left out of Wulff's notes, or it could be a simple written or publishing error. The description was so beautiful however, that I have maintained his exact wording.

Cystopteris (com.) : could be *Cystopteris fragilis*; Brittle bladder-fern, Common fragile fern, or *Cystopteris montana*; Mountain bladderfern.

Drabae : a genus of grasses more commonly known as Whitlow grasses.

Draba hirta : also known as *Draba glabella Pursh*; Smooth draba, Smooth whitlow grass.

Draba nivalis : Yellow arctic whitlow grass, Snowy whitlow grass.

Dryas : could be *Dryas octapetala*; Mountain avens or *Dryas integrifolia*; Entireleaf mountain-avens, White mountain-avens, Northern white mountain avens, Mountain avens.

Epilobium latifolium : now known as *Chamerion latifolium*; Dwarf fireweed, River Beauty, River Beauty willowherb, Broad-leaved fireweed.

Eriophorum : a genus of flowering plants in the sedge family commonly known as cotton-grasses or cotton-sedges. There are several native possibilities that Wulff could have been referring to, including: *Eriophorum angustifolium*, *E. callitrix*, and *E. vaginatum*.

Heirochloa : likely *Heirochloa alpina*; Alpine sweetgrass.

Hesperis pallasii : referenced in *Greenland by the Polar Sea* with the common names of Arctic stock and Night-smelling rocket, *H. pallasii* is a synonym for the updated *Erysimum pallasii*, now more commonly known as Pallas' wallflower.

Juncus biglumis : Two-flowered rush.

Kobresia : a genus of plants in the sedge family, sometimes known as bog sedges. Wulff could be referencing *Kobresia bipartita* (now a synonym for *Kobresia simpliciuscula*); False sedge, Simple bog sedge, Simple kobresia or *Kobresia myosuroides* (now a synonym for *Kobresia bellardii*); Bellardi bog sedge, Pacific bog sedge or Pacific kobresia. Both are native species with populations in western Greenland.

Lesquerella : *Lesquerella arctica*; Arctic bladderpod.

Luzula confusa : Northern wood rush.

Luzula nivalis : Arctic woodrush, Tundra wood rush.

Lycopodium selago : now known as *Huperzia selago*; Northern firmoss, Fir clubmoss.

Myrtillus uliginosa : now known as *Vaccinium uliginosum*; Bog-blueberry, Bog-bilberry.

Oxyria : likely *Oxyria digyna*; Mountain sorrel, Alpine sorrel, Alpine-mountain sorrel.

Papaver radicatum : Arctic poppy, Rooted poppy, Yellow poppy.

Pedicularis hirsuta : Hairy Lousewort.

Poa cenisia : now known as *Poa arctica*; Arctic bluegrass.

Polystichum lonchitis : Northern holly fern, Holly fern.

Potentilla emarginata : now known as *Potentilla nana*; Arctic cinquefoil.

Potentilla nivea : Snow or Snowy cinquefoil.

Potentilla rubricaulis : Red-stemmed cinquefoil.

Pyrola uniflora : possibly Wax flower or Single wax flower.

Salix arctica : Tundra willow, Arctic willow.

Saxifraga cernua : Drooping saxifrage, Nodding saxifrage, Bulblet saxifrage.

Saxifraga oppositifolia : Purple saxifrage, Purple mountain saxifrage.

Saxifraga tricuspidata : Three-toothed saxifrage.

Silene acaulis : Moss campion.

Stellaria longipes : Long-stalked starwort, Alpine stitchwort.

Trisetum spicatum : Spike trisetum, Spike false oat, Downy oatgrass, Northern oat grass.

Wahlbergella affinis : now known as *Silene involucrata* (also *Silene wahlbergella*); Arctic or Northern catchfly.

Wahlbergella triflorum : now obsolete, but has several possible synonyms including *Melandrium triflorum*, *Lychnis triflora* and the most contemporary, *Silene sorensenis*. Each are commonly known as the Three-flowered campion.

Woodsia ilvensis : Oblong woodsia, Rusty woodsia, Rusty cliff fern.

ACKNOWLEDGEMENTS

Several of the poems in this collection were orginally published in pamphlet form by Corbel Stone Press in 2013. 'Four Inuit Songs' appeared in *Reliquiæ Volume One* (2013) and three poems from 'Across the Inland-Ice' appeared in *Reliquiæ Volume Two* (2014). Poems from 'Across the Inland-Ice' appeared in both *Contemporary Verse 2* (2013) and *Freefall Magazine* (2014). 'Four Inuit Songs' was also published by *Contemporary Verse 2* (2014). I would like to thank the editors of these publications for their support.

By the same author

Books

An Almost-Gone Radiance (2018)
Heart of Winter (2016)
Memorious Earth (2015)*
Field Notes, VOLUME ONE (2012)*

Booklets

An Altar of Sunlight (2015)
Across the Inland-Ice (2013)
The Last Diary Entries
of Dr. Thorild Wulff (2013)
A List of Probable Flora (2013)*
Wolfhou (2013)*
Relics (2013)*
Traces (2012)
Boreal (2012)
Bearer (2012)
Wolf Notes (2010, 2011)*
Induviæ (2010)
Typography of the Shore (2009)*

* with Richard Skelton

Lightning Source UK Ltd.
Milton Keynes UK
UKHW02f0615130818
327146UK00006B/171/P